The Evolution of Hands and Feet

The Evolution of Hands and Feet

Robert Macdougall, Henry Bernard & *al.*

Part I

The Evolution of the Human Hand[1]

The succession of organic modifications which resulted in the formation of the human hand is part of the general process of evolution by which in the animal series the means of progression and of the taking of food were shaped by the environmental conditions under which life was carried on. Antecedent to the appearance of vertebrate limbs a series of manifold devices had originated by which the body could be transported from place to place and appropriate foodstuffs seized and carried to the mouth. These

[1] By Robert Macdougall.

consisted of more or less permanent extensions of the body substances, naked or clothed in protective shields of denser material. In some types the limbs were created in the act of extension itself and were retracted by absorption and disappearance into the general body mass; in some they were formed of erectile tissues which could be protracted or withdrawn as occasion demanded; in some the whole body was thus contractile, and alternately elongated and shortened as the animal progressed; in some the organs of locomotion consisted of definitely formed limbs, which, while subject to loss by violence or even sudden shock, might be repeatedly and perfectly regenerated in the course of the individual life. In the forms to which they are molded and the mechanical principles upon which they depend, these organs of movement present the utmost variety, including ameboid extensions, flagellate cilia, pulsating bells, contractile stalks and bodies, suckered tentacles, swimming fins and tails, wings and articulated legs. They appear as a great series of adaptive levels through which the evolution of this particular mechanism passed toward more highly integrated and developed types.

The functions of life which call into service the bodily limbs are chiefly two—locomotion, an activity which has arisen in connection with the search for food and flight from enemies; and prehension, which is concerned primarily with the grasping and tearing of food, but secondarily also with processes assistive of locomotion and other biological functions, such as sexual congress, the care of the body, burrowing and climbing. Of these two functions, if we regard the vertebrate class only, the former is the more primitive. Upon the office of locomotion the prehensive and manipulative activities of the limb have been superposed as subsequent and more specialized adaptations. In vertebrates of less modified types the food is seized and manipulated by the mouth parts directly. Fish, reptiles and birds feed in this way. In these as well as in mammalian forms which present relatively slight limb specialization, the mouth parts have in many cases undergone modifications which render them effective instruments for grasping, rending, digging, picking and the like. Such adaptations are shown in the snouts of the mullet and pig; the beaks of the paddle-fish, the duck-billed otter, the humming-bird and the secretary; the tusks of the boar, the horn of the rhinoceros, the proboscis of the tapir, the tongue of the chameleon and the trunk of the elephant. In all these cases the specialization of the limbs which accompanied such modifications of the mouth parts consists in an adaptation of the function of locomotion in connection with the particular conditions under which the life of the species is carried on; in consequence of which their general features have diverged very widely from those of manipulative organs.

The earliest form of locomotion which vertebrate limbs fulfilled was propulsion through the water. The problem to be solved did not include the support of the body, which was buoyed up by the dense medium in which the animal moved. The same dense medium afforded a sufficient resistance to allow of a relatively slow and weak movement on the part of the locomotive organs. The earliest vertebrate limbs, or the body extensions which foreshadowed them in times still earlier, needed neither the strength and rigidity of the terrestrial leg nor the expanse and velocity of stroke of the aerial wing.

If we conceive the progenitor of the limbed vertebrate to have progressed by means of an undulatory motion of the whole body, brought about by a peristaltic wave of contraction passing from front to rear of the animal, it is not difficult to infer the advantages which would accrue to those individuals in which a modification appeared in the form of flexible extension parallel to the longitudinal axis of the body, by the independent undulations of which progression became possible. The economy resulting from reduction of movement in the whole body mass would be accompanied by a decrease in the likelihood of attracting notice, a greater control of movements in taking food, and a more exact process of perception in adjusting the body to surrounding changes.

Though the series of limb forms is obscure in its earlier parts, the whole group is generally supposed to have its prototype in the lateral fold of the primitive fishes, in which locomotion took place through a wave-like movement passing backward along the length of the web. Out of this primitive lateral fold the various fin-formed limbs which characterize the aquatic progenitors of the land vertebrates arose by a series of modifications in which the following stages may be noted: In the undifferentiated swimming folds first developed a system of parallel rods extending from the body surface to the margin of the web, which probably both served the purpose of increasing the resistance of the locomotive organ and was accompanied by muscular and nervous developments which allowed greater definition and force in the reactions produced. Among these rods certain members outgrew the rest, a development which from mechanical causes alone would tend to survive in a bilateral form. The number of such points of origin of increased growth in the rods was finally reduced to two on each side of the body, after a series of forms which we may conceive to have presented a diminishing series of rods, as the lamprey and shark present numbers of gill arches intermediate between those of the lancelet and the perch. With the definition of these fore and hind pairs of axial spines a concomitant modification of the adjacent members of the system of parallel rods took place, in consequence of which, first, a differentiation in size arose among them, those in proximity to the axial spine increasing, those remote from it decreasing in length; secondly, changes in the points of their attachment to the body occurred, the system of secondary rods moving from the median regions in either direction toward the axial spines; and finally, these accessory rods arranged themselves in a radial relation to the central rib, thus giving anterior and posterior fan-like extensions connected by the remnants of the degenerating fold and rods in the intermediate body regions.

Further differentiation of the axial and neighboring spines, in which the latter were progressively affiliated upon the former and there appeared a definite point of articulation of the whole system with the body mass, gave rise to the bipinnate fin, a roughly symmetrical organ in which the main spine occupies a central position and is flanked by a group of supplementary rods on either side. From this form structural modification proceeded, first, by the reduction and disappearance of the accessory spines on one side of the main axis, giving the unilateral fin; and secondly, through a similar degeneration of those on the remaining side, by which the limb was reduced to a prong-like form represented in the lepidosiren. The limbs at this stage of development were in a condition which in general was more adapted to progression upon land than through the water, since the expansion upon which their propulsive action depended had ceased to be an element of importance, and all that was needed for terrestrial locomotion of a crude sort was a condition of sufficient rigidity in the limbs to allow of their use in dragging or pushing the body along, as the turtle does, but not necessarily of supporting its full weight as do the common quadrupeds. Before this final stage was reached, however, the animal had begun to practise land travel, using fins which were in the bipinnate condition as terrestrial limbs, as is the case with the Australian salmon, ceratodus.

From this primitive terrestrial vetebrate limb, through a series of cleavages of, or buddings from, its extremity, giving successively two-, three-and four-toed forms, arose finally the five-toed generalized type of mammalian limb. The subsequent modifications of this organ, if we omit the divergent series of adaptations which gave rise to the pterosaurians and finally to the birds, present forms of specialization connected with the following modes of progression, namely, swimming, running, leaping and climbing. The first, exhibited in different degrees by the whale and the dolphin, we may pass by, both because it follows a process of adaptation unlike that of the group of animals to which man belongs, and because the change may be regarded as degenerative, inasmuch as the animal returns to a medium which makes less demand upon the structural resistance of the organism than did that which was relinquished. Adaptation to running finds its extreme form in the hoofed animals, in which the body is poised upon the extremity of the limbs, thereby conserving their full length for the purpose of rapid movement by employing the utmost length of stride; and in which the number of functioning toes is progressively reduced until, as in the horse, only a single massive and horn-shod central digit forms the body of the so-called foot. In adaptation to leaping, which is presented both by animals which have passed through an arboreal stage, as the kangaroo, and by others which have always been terrestrial, like the hare and jerboa, the structural modifications consist primarily in an increase in the size of and strength of the posterior limbs, with a concomitant degeneration of the fore limbs as they are less and less called upon to share in the function of supporting the body. Along with this primary modification goes a greater or less degree of specialization in the extremities of the limb, by which, as in the case of the running animals, one or more of these take upon themselves the chief support of the body and the rest suffer functional atrophy. In the jerboa, for example, one toe only is thus degenerate, while in the kangaroo three are rudimentary.

It is with the modification of the five-toed limb for the purpose of climbing that we are here especially concerned, since it is in the arboreal group of animals only that the specialization of the fore limb in the form of a hand appears, and since it is to the adaptations fostered by this form of existence that man owes the early development of his own dextrous and accomplished manipulative organ. This modification consists, first, in the modeling of the extremities of the limbs to a form which made the act of grasping possible; secondly, in the separation of the whole system of limb terminations into two opposable groups, by which primarily a more perfect grasp was secured, and later the refined manipulation of objects was made possible; and finally, in the differentiation of hind and fore limbs, by which the former were made to provide secure and rapid locomotion and the latter were left free for specialization controlled by the sole condition of prehension and manipulation.

The first of these functions appears to be essentially connected with the habit of walking[2] on the sole of the foot — plantigrade locomotion — and not on the knuckles or toes — digitigrade locomotion. Another method of climbing exists which is common to the rodents and the cats. In these animals the act of climbing depends upon the development of claws sufficiently long, strong and sharp to be attached like hooks to the roughened surfaces upon which the animal climbs and thus to support the body. In such forms the modification is

[2] See Part III : *The Physiology of the Feet*.

of a superficial feature of the body structure and is probably a secondary function, the claws having been developed in connection with habits of seizing prey rather than of climbing. There is here no essential modification in the anatomical relations of the various parts of the limb, and it is inconceivable that any such subsequent development should be connected with this form of climbing organ as is presented in the limbs of the anthropoid apes and man.

In the plantigrade animal, on the other hand, the disposition of the limb is such that when the weight of the body is thrown upon it the toes tend to be thrust apart even when the foot is resting on a flat surface, and to be forced into a concave shape when pressed upon a rounded object. It is probable that a fair degree of development in the joints of the limbs had taken place in both flexion and separation before they were used for the purpose of climbing. But flexibility and separability of the digits form only the intial step in the process by which adaptation to an arboreal life was perfected. The second—and beyond all other changes important—modification consisted in the structural opposition of one digit to the remaining group. This differentiation occurs also in the lizards, *e. g.*, the chameleon, and in the birds, under similar conditions of climbing and perching; but in connection with such specialization of the limbs in other regards and such modification of the body system as a whole that important service in the evolution of intelligence was precluded.

In the production of opposition changes took place in the hind limbs first and most generally, since all species in which the thumb is opposed possess opposable great toes also—except in the single case of man—while many species occur in which opposition is presented by the hind limbs alone. In this adaptation of the foot to climbing three structural changes were effected—the parts of the limb became more flexible, the joints more widely separable, and the great toe, as has been said, opposed to the group formed by the remaining digits. All these are important features in rendering the limb a more efficient tool.

For the development of those peculiar functions which characterize the human hand, however, a further change in the use of the fore limb was necessary, by which it was relieved from participation in the support of the body and in primary locomotion. This relief must have taken place by a process which involved simultaneous changes in both fore and hind limbs. The support of the body, hitherto laid upon all four limbs, could not have been taken over at once in its final adequacy and security by the legs alone, unless we conceive of a spontaneous variation of improbably large extent. The animal at first raised itself hesitatingly upon its hind limbs, supporting its weight in part by the grasp of the hands upon higher portions of the trunk and branches, thus distributing the function as heretofore among the whole set of limbs, but in such a way that the fore limbs were adapted to their new specific use while performing their old generic function. The body, in this stage of development, was sustained in part by support from beneath and in part by suspension from above. Either of these factors may be conceived as appropriating a chief place in the locomotive function; and in different animal species these divergent directions of development are both presented, progression by swinging from limb to limb in the long-armed apes, and by the sole use of the legs in man.

It is probable that the progenitor of man, together with the whole group of anthropoid apes to which he belongs, maintained the quadrupedal position longer than those types which, like the Cebidæ, *e. g.,* the tee-tees and Capuchin monkeys, present no opposition in the members of the fore limbs. If we conceive the semi-upright position to have been assumed at a time anterior to the development of opposition in the hind limb, say at the beginnings of arboreal existence, so that from the outset each pair of limbs was modified under different conditions of function, it will be found difficult to imagine the causes which under these unlike circumstances brought about a similar modification in each set of limbs. If, on the other hand, both fore and hind limbs were used to support the animal in a quadrupedal position upon the branch beneath it during the early period of arboreal life, it will be as difficult to imagine a reason why both sets of limbs should not present the same type of adaptation. The condition which predisposes to conservation of the phenomenon of opposition is support, not suspension; it is peculiarly a modification of the foot. All that is involved in successful adaptation to the function of suspension is the existence of sufficient elongation in the digits, flexibility in the joints and strength in the muscles — the development of a strong and supple member, but not necessarily one possessing an opposable thumb. Even a single series of joints may form an efficient instrument of suspension, as in the case of the prehensile tail of the monkey tribe. For support upon the rounded branch beneath, on the other hand, some sort of forking is almost the only modification which could give security, and in the man-like ape this has taken the form of an opposition between a single member and the rest of the group.

We may therefore conceive that the progenitors of the Capuchins and other parallel fingered species soon after their adoption of the arboreal habit—or at least before the appearance of any important modification of the earlier structural relations of their limbs—took to a form of locomotion in which the body was partly supported from beneath by the hind limbs and partly steadied or suspended from above by the grasp of the fore limbs; so that the peculiar modification which the arboreal form of life contributed to the animal type was incorporated in the hind limbs alone. The anthropoid apes, on the contrary, which show this specialization in fore as well as hind limbs, we shall conceive to have persisted in the quadrupedal habit during a period the continuance of which was sufficiently protracted to allow of the appearance of similar modifications in all four limbs. Only subsequent to this process of adaptation should we imagine the progenitor of man to have arisen from the quadrupedal position and to have used the fore limbs for the secondary support of the body by grasping the upper branches.

In this new function the limb specialized by opposition had probably little advantage over the more primitive hand of the monkey, in so far as suspensional support was concerned. In respect to those other uses upon which the subsequent development of man in all kinds of mechanical skill depends, this new structural variation was of the highest significance. The monkey tribe gave up the habit of walking on all-fours too early and is suffering from the consequences to the present day.

This stage of development, however, represents a condition in which the factors of further evolution are confused and the various parts of the organism imperfectly adapted to the functions they are hereafter to perform. Hands and feet conform to the same architectural type. Both share in the unitary process of locomotion; the hands are capable of supporting the fore part of the body in moving, the feet are still prehensile organs. There is no exclusive functional specialization by which fore and hind limbs may be set off from each other. This subdivision of labor must come about through a development of the lower limbs by which they become capable of the sole support of the body at rest and in progress. In other words, the hands can not be released from their office of steadying and supporting the body until sufficient skeletal changes and muscular growth have taken place in the lower limbs to enable them to carry on the function of locomotion alone. The freeing of the hand for exclusively manipulative purposes thus depends upon the replacement of the semi-erect posture by a fully erect one, in which process the calf develops, the joints are straightened and the whole limb rotates upon its point of attachment to the body until the main axes of the two are parallel and each is vertical in position.

These changes could hardly have taken place during the continuance of an arboreal habit of life. The means of support afforded by the branches is too precarious, the form of locomotion which practical conditions impose upon the animal too restricted and interrupted to make the development of such a limb as the human leg possible. The need of supplementary support to which an unstable balance must give rise and the facility with which the arms can come to the aid of the legs as the animal makes its way from tree to tree are likewise factors which retard the development of efficient bipedal locomotion. The freeing of the hands may therefore be regarded as a concomitant of the return of man's progenitor to a terrestrial habitat, in which free, large and continuous movements of locomotion were both possible and necessary. Only on the wide, open spaces of the ground can we conceive the apeman to have become a swift and sustained runner, holding the body upright and the arms free.

At the same time with the changes in the leg already described the habit of traveling over the level surface of the ground would tend to produce a closer knitting of the ligaments of the foot and a greater compactness and rigidity in its general structure. The opposition of the great toe, no longer necessary to preserve the animal's equilibrium—since this is sufficiently secured in a lateral direction by the relation of the two legs—becomes a distinct impediment to land travel, owing to its interference with the movements of the fellow limb and its liability to injury by striking upon the objects among which the animal walks. With the further development of the foot, however, whether degenerative or other, we have not here to do.

As regards the special causes which led to the adoption of a terrestrial habitat in preference to the earlier arboreal life, it is probable that the change was intimately related to the development of the opposable thumb. The platyrrhine monkeys have the same type of foot as that possessed by the man-ape and do not progress predominately by swinging as do the long-armed apes. Anatomically, therefore, they differ from the progenitor of man chiefly in the fact that, unlike him, they have retained the parallel-fingered hand. In this differential feature resides their disability. The monkey form of hand is adequate for seizing and clinging to branches, but deficient in adaptability to all other mechanical purposes. For grasping and pulling, for digging and tearing, for handling stones and sticks the human hand with its opposable thumb is incomparably superior. Among the uses for which, in virtue of these capacities, it is especially fitted are the employment of weapons, the construction of means of defense from attacks by carnivorous beasts and later the use of tools.

The relinquishment of an arboreal habit involved the giving up of an important refuge and the assumption of a mode of life assailed by many new and grave dangers. The tree is a place of safety; it affords a secure retreat from some enemies and concealment from many others. A life amid its branches is compatible with a condition of weakness or defenselessness which would be fatal to the species under the circumstances of a ground habitat. To descend from the trees and venture that mode of life implies one of three possible resources: the animal must either he Heel of foot enough to distance his pursuers, or he must possess weapons of defense sufficient to repel attack successfully, or, finally, he must supply deficiencies in these regards through a cunning which enables him to escape his enemies by artifice. The apes are not swift of foot as compared with beasts of prey. They are poorly provided with natural weapons or means of defense. They have neither tusks nor claws, neither hoofs nor horns, neither great mass and strength nor impenetrable hides. If they are to take the aggressive or even to repel attack successfully it must be by the invention of artificial weapons whereby their deficiencies are made good; but as recourse to such instruments is a purely mental resort to obviate actual physical difficulties, it may be said that the ape-man met his difficulties in only one way, namely by cunning—escaping his enemy by retreat to strongholds of his own devising; meeting him, when battle was unavoidable, not with bare hands but with weapons, and taking his prey by traps and snares. But the schemes of his cunning brain could become practicable only as the result of a distinct mechanical constructiveness. Stones must be gathered and dropped or thrown with accuracy; clubs must be selected and wielded; traps must be put together after they have been devised. In all this the manipulative hand is essentially linked with the resourceful mind. With any other known type of limb the problem would have been insoluble. The specialization of the hand, therefore, with its opposable thumb and its wonderful adaptability to mechanical uses we may conclude to have been the single indispensible condition, so far as regards gross anatomical features, which determined the widely divergent subsequent fortunes of the monkey tribe and man-ape respectively.

For the principle of separation between this type and the rest of the anthropoid apes we must look to the different directions of development taken by the central nervous system in the two cases. Henceforth no important structural changes are to occur in the general features of the hand. Development is to take place chiefly through an increase in the facility and precision with which a variety of relatively simple movements are made, and the substitution, in ever increasing grades of complexity, of mechanical instruments for the use of the hand itself as a manipulative and constructive agent.

Part II

Feet and Hands[3]

I

In the great family of the backboned animals, to which we ourselves belong, many different kinds of feet and hands are to be found, their shape showing a very wonderful relation to the manner of life of their possessors.

We are not going here to describe the fins of fishes, although many people believe that our feet and hands were developed from fins; we shall only deal with the true feet and hands found in animals higher than the fishes.

---

[3] By Henry Bernard.

When we try to picture the foot or hand of any animal, we naturally think of it as we see it, often covered with fur which to a great extent hides its real shape. In our own feet and hands the true shape is more evident because the skin has lost its hairy covering. To obtain a clear idea of the feet and hands of different animals it is therefore better to limit ourselves to the bones which give them their firmness.

However different feet and hands may be, there is a certain remarkable similarity of plan in them all. In all, as in our own, there is a cluster of bones forming an "ankle" or a "wrist," and then running out from this a certain number of what we may call "rays." In each foot and hand there were once five rays, as there now are in our own, but in many animals this original number has been reduced, as we shall see. Each ray consists of several joints; the first joints are the longest, and are bound together by skin and flesh to form the sole of the foot or the palm of the hand, while the other joints form the free toes and fingers, the skin at their tips carrying nails or claws.

We can all trace this general plan both in our own feet and in our own hands, although in outward form our feet do not resemble our hands, and the work they have to do for us is so very unlike that it may seem surprising they should be built upon the same plan.

The explanation of this similarity of plan is very simple. Once, very long ago, there was no difference in the work done by feet and hands; they were all used for walking, and were thus all, properly speaking, feet. So if we want to understand our feet and hands we must go back to animals which had four feet and no hands, the four feet each having five toes.

Fig. 1 is a drawing of the foot of a crocodile, which is probably not unlike the earliest kind of foot. At the end of the leg bones a and b we see several small bones, and, starting from these, five jointed rays which in the natural position lie flat on the ground. The fore feet and hind feet of the crocodile are almost exactly alike, having the same work to do — i.e., helping the animal to shuffle along the ground. This kind of foot is very well suited for reptiles, such as crocodiles and lizards, which lead a more or less lazy life, merely moving from place to place to find a patch of sunshine to lie in, or a spot which the animals or insects they feed on frequent, and where they can be snapped at easily. Some of these animals, it is true, are capable of darting at times with lightning speed, but this they seem to do by the help of their tails.

The clumsy, shuffling way of walking on the flat of the foot has been given up by most animals, but we still find it in some, such as the bears, which are called plantigrades or sole-walkers because of their flat-soled feet. The feet of the bear are very superior to those of the crocodile, for they are armed with claws which help him to hold prey and to climb trees; but his awkward, shuffling gait shows pretty plainly that the method of moving on the flat sole is not the best possible for running.

A better method of running is found in most of the higher animals, which no longer touch the ground with the sole of the foot, but only with the toes or digits, and are therefore called digitigrades or toe-walkers. Fig. 2 is the foot of a doglike animal (the wolf), and Fig. 3 is that of a catlike animal (the lion), and in both of these we see that the part of the foot which in ourselves we call the heel does not lie on the ground as in the crocodile or the bear, but is raised high above it, so that the animal walks, as we have said, on its toes or digits. The reason why this is a better method of running is that it gives extra length to the leg. A longer leg means a longer stride and greater speed. The wolves, hyenas, and foxes, which are doglike animals, and the lion, leopard, lynx, and all other catlike animals are beasts of prey—that is, they eat other animals which they have to hunt—and all in their turn are liable to be hunted, so they need to be able to run very swiftly. Many of the catlike animals, too, depend, both in attack and defense, quite as much on springing as on running, and so need to be very agile. For both running and springing we ourselves even use our toes and not the soles of our feet, and these animals, with whom running and springing have become a constant habit, have come to use only their toes always;

the heel no longer touches the ground except when the animal is crouching.

Now, in the foot of the crocodile (Fig.1) it is seen that all the toes are not of the same length. If animals with such feet began to walk on their toes, some would not touch the ground. This, indeed, is what we find in many animals, especially in those cases where running is all that is required. In the feet of the dog, for instance, only four toes touch the ground; the fifth toe, having long been less used than the others, has become very small. In the hind feet, which are used exclusively for running, the remains of the fifth toe can be felt as a small projection under the skin, some way up the back of the leg. In the fore foot of the dog, however, it is a distinct toe, with a nail, which is still used for digging.

FIG. 1.

FIG. 2.

FIG. 3.

FIG. 4.

In cats and dogs, as in the bears, claws are well developed. In the dogs they are short and hard but not very sharp, and are used chiefly for digging. The catlike animals, on the contrary, have very sharp, hooklike claws, which are of great use in seizing and holding prey.

It is obvious that if an animal, changing still further its method of walking, took to running on the tips of its nails or claws instead of on its toes, it would have a still greater length of leg, and therefore would be still better fitted for running swiftly. This remarkable manner of running is actually found in most of the swiftest four-footed animals, such as the horses and deer, and also in the cows, the sheep, and the pigs, which are therefore called ungulates, *ungula* meaning a nail. All these animals feed chiefly on grass, of which they need great quantities to nourish their usually large bodies. To obtain constant supplies of grass, in a wild condition, they have continually to roam from place to place. Their feet are admirably suited for these roaming habits, and also for very swift running, which is their best chance of escape from their enemies, the flesh-eating wolves, tigers, etc.

The foot of the bison (Fig. 4), in which the tips of only two toes touch, the ground, shows very clearly the position assumed by the bones of the foot in these animals. Not only are the heel and the first long joints raised above the ground, as in the dogs and cats, but the two remaining long upper joints join together to form one strong bone (Fig. 4), and none of the lower joints of the toes touch, the ground. The nails of two toes which form what we call the "hoof" rest upon it, while at *a* we see all that is left of a third and fourth toe, possessed by the distant ancestors of the bison.

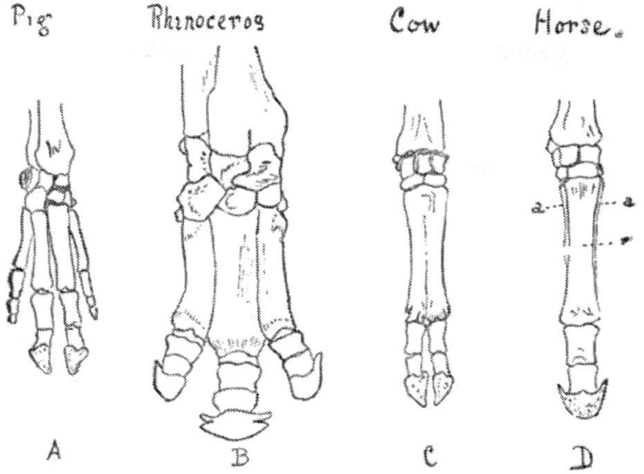

FIG. 5.

The chief distinction between the feet of the different nail-walkers is the number of toes whose nails form the hoof. Fig. 5 A shows us that the pig has four toes, but only two touch the ground and make the hoof; the other two are useless, and are gradually becoming smaller. The rhinoceros (Fig. 5 B) has only three toes, and these are not all equal in length, but all are in use and end in massive hoofs. The cow (Fig. 5 C) has only two toes, the upper joints of which have grown together into one as in Fig. 4 (for the bison is one of the cows) so as to form a long, strong part of the leg, the two very thick nails making a double hoof. This is what is called in the Bible a "cloven" hoof; it is not, however, as was supposed, a single nail split down, but two distinct nails belonging to two toes. Then, lastly, the horse (Fig. 5 D) has lost all its toes except one, which has become exceedingly thick and strong. The horse on its four feet, each, ending in only one toe with its great nail, attains immense speed in running, great length being given to its stride by the lengthening out of the one bone (*), which remains as the representative of the "sole" of the foot. Ages ago, as is proved by fossil remains, the forefathers of the horse of to-day had three, four, or even five toes. At *a, a,* on each side of the great single bone two thin bones can be seen. These two bones are the splint bones, the only remains now to be found of the vanished toes.

All the animals we have so far mentioned have four feet more or less alike, all being used for the same purpose, that of running.

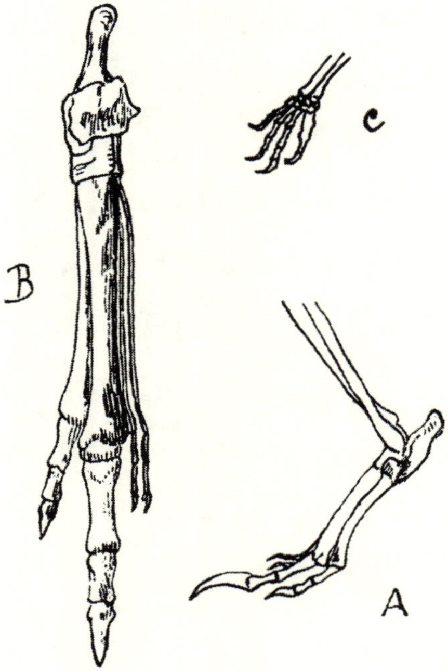

FIG. 6.

But some animals came to use their fore feet for one purpose and their hind feet for another, and in consequence of this the fore feet came to be unlike the hind feet.

A striking example of such a difference in the use of the fore and hind feet, leading also to a difference in their structure, is found in the kangaroo, an animal which is seen wild only in Australia, where it hops or leaps over the open country, more or less upright, with extraordinary swiftness by means of its hind feet alone. When it rests, it sits on the long soles of its hind feet, steadied by its thick tail. Fig. 6 A represents the hind foot of a kangaroo in its ordinary position when leaping, and Fig. 6 B shows of what strangely changed bones it is composed. One ray has become very long and thick, and another, though not so long, is also fairly thick, but the other two are quite thin, as if they were dying away from not being used.

The fore feet of the kangaroo (Fig. 6 C) are never used as walking feet except when the animal is hobbling about slowly. Their chief work is grasping and tearing the leaves, grass, or fruit, or digging up the roots which form the food of the kangaroo. They are never needed to support the weight of the body, and so the toes are not large and thick, and four of the five toes are kept. In looking at the fore feet of the kangaroo one is tempted to call them "hands," for this very interesting difference in the use of fore and hind limbs in other animals, such as the monkeys, gave rise gradually to a true grasping hand.

In the second section we shall deal with the extraordinary transformations of the fore feet into paddles and wings, found in whales, bats, and birds, and shall also see how true hands came to be developed in monkeys and in man.

II

In the previous section we described the feet of some of the chief groups of four-footed animals. We saw that in most of these animals the four feet are very much alike, because they all have the same kind of work to do—that is, walking or running. But when, in consequence of its manner of life, an animal comes to use its fore feet differently from its hind feet, as the kangaroo does, we find that a difference arises in the structure of the two. We now have to trace some of the more marked changes of this kind.

So far, all the animals we have mentioned have been land animals, all needing their feet for moving in one way or another on the ground. But, ages ago, some land animals changed their manner of life and took to living chiefly in the water, only occasionally coming on to the land, or even merely coming to the surface of the water when they were obliged to breathe, for they still had lungs, which needed to breathe air. To such animals (for example, seals and whales) ordinary feet were partially or altogether useless; they had chiefly to paddle or swim, seldom or never to walk, and, to enable them to hold their own against their water enemies, their limbs gradually became very much changed.

Taking first the seal, which even now climbs on to the land, we see all the four feet changed into paddles, but these paddles, being still occasionally needed for use on land, are not very unlike the feet of an ordinary land animal. In Fig. 1 we see that the fore foot of the seal (A) bears a considerable resemblance to his hind foot (B); there is some difference in length and in the position on the body, for the hind limbs face each other at the back of the body, but all four are paddles.

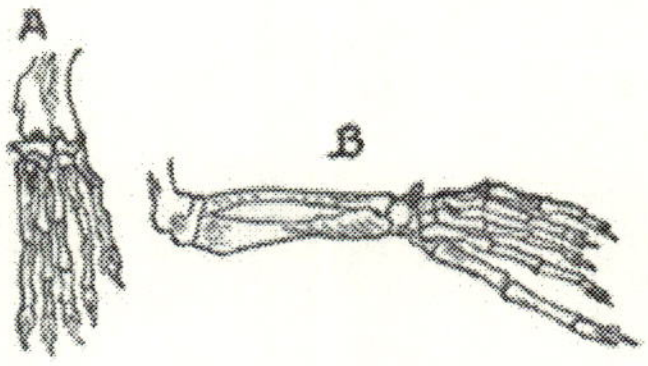

FIG. 1.FIG. 2.

Taking next the whale, which has entirely given up coming on to the land, we find only one pair of paddles or changed feet. When the land ancestors of the whale took to living in the water, they probably had tails, which, by whipping the water, helped them to swim, and they evidently learned to wave the hinder part of the body in the water as a fish does to help to push him along. They did not use their hind limbs at all for this purpose, and these, like all unused parts of the body, diminished in size and strength. In the whales now existing no hind limbs are to be seen, but traces of the skeleton of the hind limbs are still found imbedded in the sides of the fishlike body. How small these remains of the former leg are in comparison with the constantly used front foot, now a paddle, may be seen from Fig. 2 A and B.

Again, ages ago, other land animals managed to raise themselves from the ground into the air, the better, no doubt, on the one hand, to escape from pursuing enemies, and, on the other, to catch flying insects. For supporting themselves in the air, wings of some sort were necessary, and so the front legs in such animals gradually altered.

Although none of these first flying animals are now in existence, we know pretty well what they were like, as impressions of their bodies have been left upon slatelike rocks. They were animals in many ways like reptiles—i. e., belonging to the same group as the crocodile, whose five-toed fore foot was illustrated in our first paper.

In some of these first fliers all the five toes were retained; four were of ordinary length, but the fifth was immensely long—longer, in fact, than the whole body of the animal—and between it and the body a skin was stretched, making a wing.

Here we have somewhat the same method of flying as we now find in the bat. In a bat's outspread wing (Fig. 3 A) the five rays of the fore foot are quite clear. Four of them are joined together by thin skin which stretches back to the leg and the tail. The lower part of the arm and three of the rays which carry the flying skin are greatly lengthened, and so a very large wing is obtained. The first ray, which we might call the thumb, remains short and ends in a strong claw that is of use in climbing.

We thus, in the bat, have an animal with the fore foot extraordinarily changed to suit it for a special manner of life—i. e., for flying. The hind foot (Fig. 3 B), not having any new kind of work to do, is much less changed. It has five ordinary toes ending in claws, with which the bat climbs or hangs on to trees. So well fitted are the hind feet for this kind of work that many bats always sleep wrapped up in their wings as in a mantle, hanging on to the branch of a tree, head downward.

But not all of the first flying reptiles developed wings of stretched skin. Some came to fly by means of feathered wings not unlike those of birds. Fig. 4 shows the bones of the fore foot or wing of a small creature called the *Archæopteryx*, or first feathered flier. This animal had three distinct toes, each ending in a claw. Two toes had thus already been lost, the fourth and fifth, which also are never found in any bird. The feathers of this creature grew out of the skin of the arm and of the third finger.

In the fowl (Fig. 5) and all other birds the wing is, as in the *Archæopteryx,* a feathered fore leg and foot, a, b, c, in Fig.5, are the leg bones, and the rest represents the foot, which retains in part the three rays seen in the *Archæopteryx.* Two of these rays, however (1 and 3), instead of having several joints, are nothing more than short, pointed bones, U C and R C are the only bones of the "wrist" left, and MC II and MC III show all that remains of the bones which in a four-footed sole-walker form the flat of the foot. 1 is the remains of the first toe or digit, 2 is the long second digit, which still has three joints, and 3 is the one remaining joint of the third digit, which is not distinctly separated from the second. All the other parts of the lizardlike fore foot have disappeared, because this was quite enough to serve as a rigid support for the skin which carries the feathers.

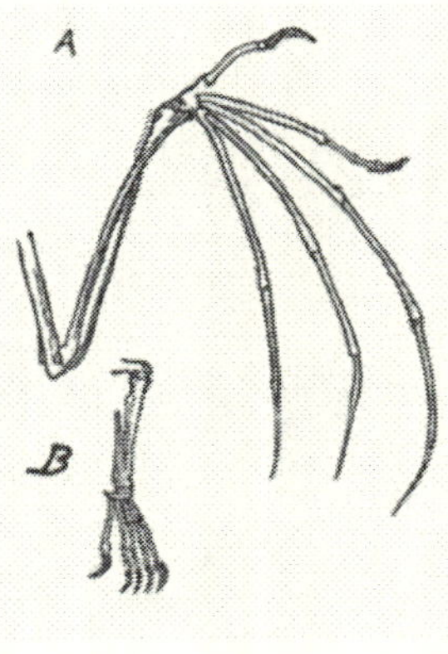

FIG. 3.

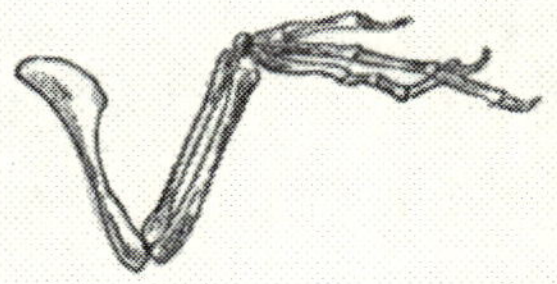

FIG. 4.

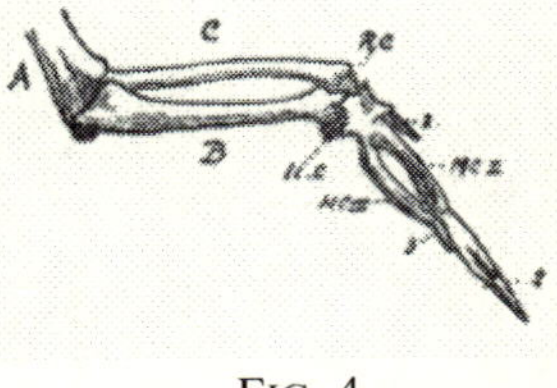

FIG. 5.

The domestic fowl, of course, uses its wings very little; it walks more than it flies. Most birds use their wings much more, and some of these have more of the bones of the original foot left. Other birds fly still less than the fowl. The penguin, for example, uses its wings not for flying but for swimming, and the bones have become much flatter and broader than those of ordinary birds. The feathers that cover them are exceedingly small, almost more like scales than feathers, so that the whole wing is flat and paddlelike, and is very suitable for acting as an oar in the water.

In some of the ostriches, which also never fly, the bones which support the wing are reduced to mere stumps, even the second digit having only one joint.

The hind feet of birds are less changed than the fore feet, but still differ from the lizardlike foot in the number and character of their toes. Some birds have four toes, others only three, and these toes are usually able to curve so as to grasp the branches of trees, and are armed with claws which help in climbing. In the birds which run along the ground instead of flying, like the ostrich, there are usually only two toes, which become very much thickened. Here, as in the horse, we find increased speed in running obtained by reduction of the number of toes. The horse, as we saw, actually has only one toe, and in some ostriches the second toe is so reduced that they practically have only one also.

Great as is the difference between the fore and hind limbs of the bird, that between our own hands and feet is quite as important. The fore feet of all animals have sometimes been called "hands," to mark the fact that they correspond with what in a human being is the hand, but in the animals we have considered we have had nothing like a hand in the true sense of the word. In a true hand, the inner finger is able to move in such a way as to face the other fingers, and the hand is thus able to grasp any object far more firmly than if the movement of all the fingers was similar, as it is in the foot. Such an inner finger is called an opposable thumb, and its presence is absolutely necessary to a real hand.

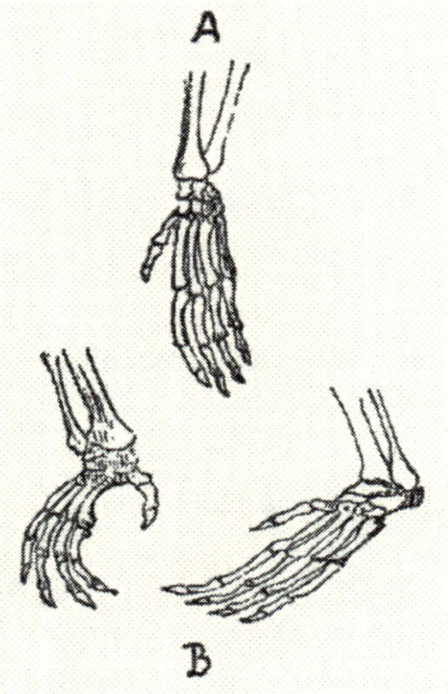

FIG. 6

Let us see if we can understand how such true hands arose. Besides the animals which run about in various ways on the ground, those that have taken to living in water and those that have learned to fly in the air, there are others that live almost entirely in trees, climbing or springing from branch to branch, and only occasionally walking on the ground. These animals, the monkeys, which feed principally on the fruits that grow on the trees they inhabit, need agility in climbing rather than swiftness in running, both for obtaining their food and for escaping from their enemies, and so all their feet have become specially adapted for firmly grasping the branches of trees—i. e., they developed thumbs on all their feet; for this reason they have been called the *Quadrumana,* or four-handed animals.

If we examine the hand and foot of a gorilla, given in Fig. 6 A and B, we shall notice that the hand (A) differs very little from the hand of man, given in Fig. 7 A. The feet of the ape, however, differ considerably from those of man (cf. Fig. 6 B with Fig. 7 B).

There is no doubt that the very earliest men spent their lives much as monkeys now do. They probably lived partly in trees before they learned how to build houses in which they could take shelter from bad weather and when threatened by animals that were much stronger and could run far more swiftly than they. Thus our hands were once developed for climbing about trees, and since that time have changed very litle in shape. As man became civilized, his hands were put to nobler uses; they were his best instruments for accomplishing all he achieved. Our hands are thus much more flexible Fig.7. and capable of a far greater variety of movements than the hands of the less civilized monkey.

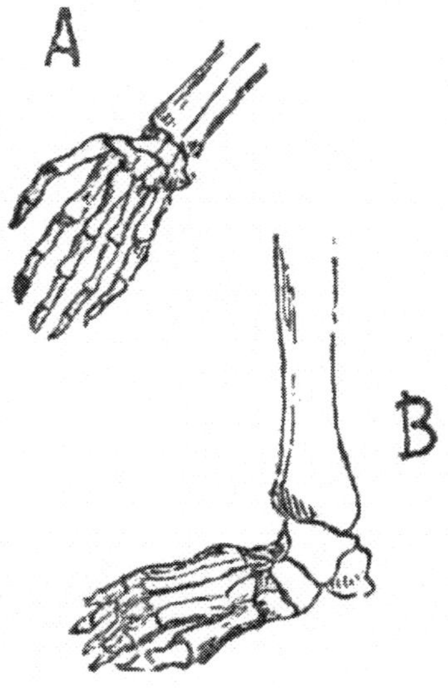

FIG. 7.

The hind foot of the monkey was not well suited for man's way of life. One great difference between man and the monkeys is that man walks and runs in an upright position on his hind feet alone, while even the most manlike of the apes walk only partly upright, not supported entirely by their feet but helped along by their hands, which, on account of the great length of the arms, can easily reach the ground. In a man, walking upright, the weight of the body is thrown entirely on the hind limbs, which in him become the lower limbs, and the feet are needed to form a firm pedestal for the whole upright body. Thus in man, the once climbing feet having given up climbing, the useless thumb gradually changed back again into a simple great toe, and the first joints of the rays lengthened out to form a firm flat sole. We can still see some signs of the former thumblike condition of the great toe in little children and in some foreign races. In babies, the great toe is often more or less opposable to the others, and among the Japanese it is still so much so that a Japanese workwoman often holds the work she wants to stretch between her toes, which are able to grasp it firmly.

Men and women have, as a rule, five toes and five fingers, the largest number found in animals; but it is by no means certain that this number will always be retained. The foot of a civilized man is always covered by stockings and shoes, and the toes are hardly used at all, and, like the toes not used in the feet of animals, appear to be diminishing in size. The little toe in a comparatively large number of human beings already has only two joints, and some of the other toes also seem to be becoming shorter. It is thus quite possible that at some future time the human little toe will altogether disappear, and that man may have only four toes, or even, in course of time, fewer than four.

Part III

The Physiology of the Feet[4].

Disregarding the action of those parts not affecting the feet, the act of walking may, as I think, be thus described: The foot put forward should reach the ground when nearly flat; the toes, the organs of feeling, should be the first to reach it, not the heel, which could not be without some concussion, however slight. The heads of the metatarsal bones and the toes are then pressed firmly against the surface. The great-toe, having only two phalanges, is held down in its whole length, the flexor tendon being attached to the final phalanx close to the joint

[4] By Thomas S. Ellis.

between them. The little toes touch at their tips only; their flexor tendons being also attached to the final phalanx of each, traction on them causes a rising at the joint between the two proximal phalanges as the tips of the toes are drawn backward. By this arrangement, in the one case a firm, solid base is formed from which the body can be propelled onward; in the other an additional hold on the surface, by a rudimentary action or grasping, is afforded. As the body is moved onward, the extensors of the great and of the little toes, without lifting them from the ground, where they are held by the flexors, draw the leg forward, while the anterior tibial, in assisting this movement, serves another purpose. It is attached to the crown of the arch, and in action tends to prevent any sinking there as the weight of the body comes upon that structure. This purpose is much more effectually served in another way: the muscles of the calf allow the heel with firmness and precision, but withal gently, to touch the ground, and the step is completed.

The heel is then raised, but the weight of the body is not borne, as commonly stated, by the muscles acting on the heel and by them only: the deeper muscles, the posterior tibial with the long flexors and the long peroneal, acting round the inner and outer side of the ankle respectively, all of them assist in raising the body and at the same time have a most important influence in maintaining the arch. The tibialis posticus, attached by its expanded tendon to the tarsus on the under surface beyond the astragalus, the bone on which the weight of the body rests, materially assists in supporting the arch from below. The long flexors passing beneath the arch from one abutment to the other are, in relation to it, as bowstrings to a bow, or rather, as the two tendons cross each other, they may better be compared to the tie-rods of a roof.

This arrangement of the two tendons crossing each other is very curious: that going to the great-toe is lowest in passing round the ankle, in order to be, as nearly as possible, at the extremity of the arch or bow at that end, as it is at the opposite one; if, however, it passed directly across the sole from end to end there would be little if any free space beneath, but, being crossed by the flexor longus digitorum, which comes round the ankle at a higher level, it is so drawn up that a hollow beneath the arch is formed; the flexor accessorius, by drawing back the tendon of the flexor longus pollicis, compels it to cross the other nearer to the heel, and so increases this effect. Thus the flexor longus pollicis, regarded as the chord of the arc, becomes itself an arc.

The tendency to inversion which all these muscles, acting from the inner side of the ankle, might occasion is corrected by the long peroneal on the outer side; it also, acting on the base of the first metatarsal bone, a point considerably beyond the center of gravity, has a bracing action on the arch, as the weight of the body falls upon it. In this, too, no doubt the small muscles of the sole assist those of the calf, but I cannot accept the converse statement that it is the "muscles of the sole assisted by the tibial muscles" which "are the active agents." The deep muscles of the calf have much the more potent influence. Thus it is that by the action of muscles the whole of the strain which the weight of the body in walking would otherwise throw on the ligaments binding the arch together is removed, and any tendency to flattening of it prevented.

This, which has been called my "bowstring theory," is the view I put forward in a little monograph, "On the Arch of the Foot," written and printed in 1877. For reasons therein given I could not accept the view that the arch is maintained by ligaments, or believe in the carriage-spring movement of those ligaments, yielding to the weight of the body, as the explanation of a springy gait. It is really due to the heel being gently lowered and firmly raised. Upon this the grace of walking depends. On the same grounds I hold that in proper walking the foot does not lengthen. Camper, whose treatise is regarded as classical, but which, as I think, contains many important errors of fact and of induction, said that his knowledge of anatomy taught him that it did so. On the contrary, I believe that as the tightening of a bow-string approximates the ends of the bow, so the bowstring action of the flexor muscles on the arch of the foot tends to shorten it. If walking were a succession of standings, flat-footed alternately on either foot, no doubt there would be lengthening, as the ligaments of the arch yielded. Such a mode of progression is, we know, possible; and, indeed, we sometimes see something like it, hardly, however, to be called walking. I would ask those who believe that the foot in walking lengthens "one tenth of its length, or about an inch" (a statement on high authority made during the past year), to consider this: What, then, would be the condition of the sole, after a long walk, from friction caused by the necessary sliding with the weight of a man borne upon it? As in every mile of the ordinary march of soldiers more than a thousand steps are taken on each foot, the result would not be pleasant even to imagine.

The position of the foot is important. To turn out the toes seems to me to be not only untrue to nature, but objectionable as well as inelegant. Camper regarded it as incontestabiy the proper position. For the following reasons I believe that the toes should be directed forward, the inner margins of the feet parallel: It is desirable that the propulsion of the body onward from, and consequent thrust backward on, the foot, and especially on the great-toe, should be in the direction of its length rather than obliquely across it, not only as giving a firmer bearing from which to propel the body onward, but as diminishing the friction on the sole and consequent tendency to foot-sore. This applies also to the smaller toes in a less degree. The long axes of all the toes continued backward seem to converge on the heel. By standing with the bare foot and springing forward it can readily be seen how much more tendency there is to slide on the sole when the foot is turned out than when it is directed forward. In the latter position, too, the arch is much more firmly braced up—a fact recognized by surgeons who advise, in cases of flat-foot, that the toes should be directed inward rather than outward. In standing, the everted position is not more stable. When a body stands on four points I know of no reason why it should stand more firmly if those points be unequally disposed. The tendency to fall forward would seem to be even increased by widening the distance between the points in front, and it is in this direction that falls most commonly occur.

Those who look on the human foot as fully partaking of the beauty of which artists in every age have regarded the human body to be the highest expression, ought not readily to admit that the boot which best conforms to its outline, reveals its features, and expresses its leading characteristics, will require an apology for want of elegance. I, at any rate, can not admit anything of the kind. The human foot is, moreover, an object of far more than the ordinary interest belonging to every part of the human structure. In the monograph already mentioned I ventured to suggest that, anatomically, there is no more marked distinction between man and the lower animals than is to be found in the special development of the foot.

However much we may regard it as in itself calling for admiration on account of its fitness for the purposes it has to fulfill and for others it may on occasion serve, the human foot is far more remarkable as an adaptation of the mammalian type, modified to suit a purpose kindred to but differing from that which the corresponding member supplies in other animals. The heel has its special form and signiticance in that man only has one adapted for crushing an offensive object beneath it. The large size and important function of the great-toe is also a specially human feature. In the mammalian typical limbs the bones of the hand and foot (or rather, to avoid confusion, in four-handed or fourfooted animals, *manus* and *pes*) are arranged on a uniform plan: to each five digits, the first having two phalanges and the others three. The first digit is generally attenuated, often suppressed, but whenever it exists it has two phalanges only.

This curious difference is nowhere, so far as I know, explained. I cannot discover that any animal (below man), recent or fossil, exists or has existed from the times of the Trias formations till now in which this arrangement has appeared to be essential. It may be of some advantage in the quadrumana, and doubtless the human hand is thus better fitted for its functions, but it seems to me to be much more difficult to imagine it possible for any other arrangement to exist in the foot unless the whole scheme of it, so to speak, were changed. It is essential that the only joint in the great-toe should be drawn to the ground by the strong flexor tendon attached to the final phalanx close to it; if another joint existed it must rise up, as occurs in the other toes, and the solid bearing would be lost. Apart from this, it must be admitted that it is mainly due to the special development of the great-toe in a line with the long axis of the foot that man is enabled to exercise the attribute, in all ages regarded as a noble one, of standing erect. Yet this special feature is the one which the conventional boot does most to conceal, and in direct proportion as it is successfully concealed the wearer is supposed to be dressed in good taste. It would seem to be regarded as necessary to reduce the foot to even-sided symmetry; but there is no law of beauty which requires this. Mr. Ruskin assuredly would not say that it is in any of "the eternal canons of loveliness" decreed that an object to be beautiful must be symmetrical. An architect required to provide more space on one than on the other side of a building would not seek to conceal or even to minimize the difference; he would seek rather to accentuate it, and give the two sides of the structure distinctive features. To me it appears that it is on this principle only that a boot, to be at once useful, graceful, and appropriate, can be designed.

Moreover, the sense of symmetry, natural and reasonable where the same function has to be performed, is, or ought to be, satisfied by the exact correspondence of the two feet, which, taken jointly, may be described as the two halves of an unequally expanded dome, irregularly extended at the base, the greatest extension being in the line of the greatest expansion of the dome, through which line the division runs. The dividing-line thus makes the margin of the two feet parallel to each other. It may be that the inner margin of the great-toe, if produced backward, would fall a little distance from the inner side of the heel. A perfect adult foot, in which the great-toe is not and never has been diverted outward, and in which there has been no consequent thickening of the large joint, is not easy to find. In children the inner line is often visibly concave. It may be remarked that in rest the great-toe is everted as well as drawn upward, in which position the toes are usually packed in a tight, medium-pointed boot; it is only therefore in action as the toes are pressed against the ground that the full extent of the approximation to a straight inner line is seen.

In the boot it is of first importance that the sole (technically, the part in front of the hollow or "waist") should allow the great-toe to occupy its natural position; it must, therefore, be straight, or nearly so, on the inner margin; but it is of little use to provide the room thus given unless it be occupied; the foot must be invited to occupy it by giving plenty of room in the upper leather on this side. It is obvious that where a flexible material is fixed on both sides and left loose between, it can be drawn farthest from the surface to which it is attached, in a line midway between the fixed parts. For this reason the highest part of the foot, which is on the inner side, will, in any case, have some tendency to go to the middle line of the boot where most room can be made; this tendency is largely and needlessly aggravated by the high ridge of the last being along the middle line instead of being on the inner margin. But not only ought the room there given to be much more according as the foot projects the more upward; it ought to be proportionally more. I have insisted that the foot does not lengthen in walking, but rather shortens. This shortening, due to the powerful action of the long flexors, causes the foot to rise across the middle, the rising being almost entirely on the inner side. Here, therefore, over (not by the side of) the ball of the great-toe is special room required. The lateral thrust, too, already spoken of as consequent on turning out the toes, tends much to displace the foot and to throw it against the outside of the boot. Toe-caps also are objectionable, as giving the most room in the middle line and inviting the great-toe to occupy that position.

No useful or ornamental purpose is served by leaving space unoccupied round the outer margin of the sole opposite the little toes, as if it were necessary to make the two sides of the boot to match. No angle existing in nature, none should be represented; the outline of the sole of the boot should conform to that of the foot.

If it be true, as already stated, that grace in walking, as well as free propulsion of the body onward, and maintenance of the arch in walking, are all due to the free action of the flexor muscles, letting the heel down with gentleness and precision, and raising it with firmness and vigor, it follows that none of these can be if the boot prevent the heel from going down, a necessary antecedent to springing up. A low heel, therefore, if any, is imperative. The perfect boot should have none. Nor can the free action of the flexor muscles have full effect, so as to draw the toes to the ground, if the sole be turned up at the toes, especially if it be a stiff one. Some turning up will come from walking, however good, but there is no reason why it should be aggravated by having the boot-last so. In it the sole should be flat to the end. Objection is made that boots with a straight, or nearly straight, inside line give the feet a pigeon-toed or inverted appearance, which is unnatural. This is not altogether due to the eye being accustomed to a more conventional pattern. It is, I am sure, mainly due to a removable cause. Any conspicuous line, that of the laces or a seam, is always carried from the front of the leg to a central point on the upper surface of the foot. The eye, falling on this line, in imagination continues or produces it, and so divides the front part diagonally into two very unequal portions, the larger on the inner side. I have found that if the line of the laces or seam be kept parallel throughout to the line of the long extensor tendon—in other words, along the crest of the ridge, thus marking out a distinctive feature—the unpleasant effect is removed.

In proper walking, which cannot be done in an improper boot, friction of the foot on the sole and of the latter on the ground is reduced to a minimum, the sole being pressed against, not rubbed along, the surface. This is shown by the very little wearing of the leather; when at last it does give way, it should be at an oval spot a little distance from the margin of the sole, under the middle joint of the great-toe. This is the last point to leave the ground in walking; here, therefore, is the greatest tendency to slide on the surface and consequent friction.

The great-toe, in any but the most perfect feet now to be found, is so easily diverted outward that socks and stockings with a straight inner line are very desirable; indeed, no others can be said to really fit the feet. When any considerable distortion exists, a separate stall for the great-toe is necessary.

For the reasons given, a last should have the inner margin nearly in a line with the inner side of the heel, and joining in front by a rounded angle a long curve on the outer margin, where no angle should be shown. The sole should be flat, touching the base-line in front. The thickest part (highest vertically) should throughout be on the inner margin, especially above the ball of the great-toe. The boots should have low heels—to be perfect, none. The line of laces or of any conspicuous seam down the front should be in a line corresponding with the inner margin of the foot along the highest part. The boot should, if possible, be left on the last for a considerable time, to overcome the tendency of the leather to recoil after the forcible stretching to which it has been subjected, and so to fix it in its proper shape.

Made in United States
Orlando, FL
04 November 2026

90205469R00062